Contents

Preface   *page* vii

   1 **Elements of Technical Writing** 1

   2 **Technical Papers** 13

   3 **Technical Letters** 26

   4 **Oral Presentations** 41

   5 **Presentation of Technical Data** 53

   6 **Statistical Analysis of Experimental Data** 82

   7 **Resumé Writing** 111

Appendix I: COMMON ERRORS IN WRITING 121

Appendix II: PUNCTUATION 123

Appendix III: COMMON WORD ERRORS 133

Appendix IV: INTERNATIONAL SYSTEM OF
PREFIXES AND UNITS.................................137

Appendix V: THE GREEK ALPHABET AND
TYPICAL USES 139

Appendix VI: STRAIGHT-LINE PLOTS FOR
SOME MATHEMATICAL FUNCTIONS.................141

References   145
Index   147

The appendices treat common errors in writing, including punctuation and commonly confused words; general information, including the international system of numerical prefixes and units and the Greek alphabet; and uses of straight lines to represent some mathematical functions.

This guide is intended for all science and engineering majors. The careful reader may notice that many of the examples are taken from the authors' experiences in materials science and engineering.

Clear communication is a challenge that often does not appeal to engineers and scientists. However, the responsibility of ethical scientists and engineers is to ensure that humanity benefits from their knowledge. If one is unable to communicate one's ideas effectively, then for all practical purposes the work is lost. Academic grades and future careers are dependent on good communication skills. Becoming a good writer is a lifelong journey, and the authors hope that this book provides a quick reference for readers in both their academic and their professional careers.

REPORTING RESULTS

1 Elements of Technical Writing

The ability to communicate clearly is the most important skill engineers and scientists can have. Their best work will be lost if it is not communicated effectively. In this chapter, elements of the technical style of writing are examined. Technical writing differs in presentation and tone from other styles of writing; these differences are described first. The most important elements of the technical writing style to be discussed are conciseness and unambiguity. The chapter ends with a discussion of proofreading and some helpful hints in developing technical writing skills.

Presentation and Tone

Technical communication differs from fiction in many ways. In mystery novels the reader is kept in suspense because the writer has hidden important clues that are explained at the end of the story to produce a surprise. In contrast, the readers of technical writing are given the important conclusions at the beginning, followed by evidence supporting those conclusions. The following example illustrates the difference. The simple question *Do we have any mail today?* can be answered by a man sitting on his porch in two ways.

He could say: "I got up out of my chair and sauntered out to the mailbox. I looked up before opening the box and saw the mailman going down the street past our house. When I opened the mailbox there was nothing in it, so I don't think we'll have any mail today."

Or he could answer: "No, we won't have mail today. The mailbox is empty and the mailman has passed our house."

Note that in the first reply, the reader must wait until the end of the story to find the answer. This is typical of fiction writing. In the second reply the answer is given up front and then justified. The tone of the second reply is kept factual. This is what technical writing should do.

Number, Voice, and Tense

Most technical communication is done in the third person. Pronouns like *you*, *I*, and *we* are to be avoided. Only Nobel laureates may write in the first person without seeming to be pompous.

Readers probably studied voice in an English class. As a reminder, examples of the different types of voice are:

Active voice: The ice melted at 0°C.

Passive voice: The ice was melted by convection heating.

Imperative voice: Place the ice in a convection oven until the ice melts.

The imperative voice is seldom used in technical communication except when giving instructions about how to do something. It tends to sound like the author is ordering the reader to do something. There is a strong temptation to overuse the passive voice in technical writing to avoid using *I* and *we*; however, it is good to use the active voice wherever possible.

Past and perfect tenses are used in technical writing, because they are used to report something that has happened. The difference in tenses is illustrated by the following:

Past tense: A break in the circuit interrupted the current.

Perfect tense: A break in the circuit has interrupted the current.

It is usually best to pick a tense and be consistent with it in your writing. Frequent shifting of tenses can leave the reader confused. Occasionally, the past perfect tense can be used to describe a prior event. The previous example written in the past perfect tense is "A break in the circuit had interrupted the current." An exception when it is okay to use the present tense is when stating an enduring truth like "Current passing through a resistor causes it to heat up."

Conciseness

A hallmark of good technical papers and reports is that they are as concise as is consistent with being complete and unambiguous. Most readers are busy people, and the writer should avoid wordiness and redundancy. In writing a technical report, one can often assume that the audience is familiar with the scientific and engineering terminology.

Consider the following excerpt from the middle of a doctoral thesis proposal.

> A schematic illustration of the spot friction welding process is shown in Figure 1. The process is applied to join the two metal sheets as shown. A rotating tool with a probe pin is first plunged into the upper sheet. When the rotating tool contacts the upper sheet, a downward force is applied. A backing plate beneath the lower sheet is used to support the downward force of the tool. The downward force and the rotational speed of the tool are maintained for an appropriate

Table 2.1. *Common journal formats for papers*

Style 1	Style 2	Style 3
Title	Title	Title
Abstract	Summary	Abstract
Introduction	Background	Introduction
Experimental Procedures	Theoretical Model	Experimental Procedures
Results	Results	Results and Discussion
Discussion	Discussion	Summary
Summary	Conclusions	Acknowledgments
Acknowledgments	Acknowledgments	References
References	References	
Appendices	Appendices	

Sometimes the nature of an article calls for a different format. A few formats from articles in different journals are listed here:

From *International Journal of Mechanical Sciences*

Title: "Influence of Strain-Path Changes on Forming Limit Diagrams of Al 6111 T4"

Format: Abstract, notation, introduction, experimental procedure, results, forming limits in plane-strain after various prestrain paths, industrial observations, conclusions, references

From *Journal of Applied Physics*

Title: "Operation of Bistable Phase-Locked Single-Electron Tunneling Logic Elements"

Format: Abstract, introduction, principle of operation, model, bistability of an isolated gate, return map, locking of a single-gate to a sinusoidal input signal, interaction between coupled gates, signal transfer in separated clock stages, circuit implementation, conclusions, acknowledgments, references

From *Journal of Aerospace Science*

Title: "Large-Eddy Simulation"

Format: Abstract, introduction, formulation, sub-grid scale models, numerical methods, achievements, challenges, conclusion, acknowledgments, references

From *AIChE Journal*

Title: "Penetration of Shear Flow into an Array of Rods Aligned with the Flow"

Format: Abstract, introduction, prior work and objective, validation of technique, results for shear-driven flow, conclusions, acknowledgments, appendices, references

From *Journal of the Astronautical Sciences*

Title: "Tracking Rigid Body Motion Using Thrusters and Momentum Wheels"

Format: Abstract, introduction, system models, dynamics, kinematics, tracking controllers, numerical examples, conclusions, appendices, acknowledgments, references

Note that all of these formats start with an abstract (or summary), followed by an introduction to the subject material, and end with conclusions, acknowledgments, and references. Appendices, if any, are at the end.

Title

Titles should be short and not too general or specific. For example, the title "Analysis and Comparison of the Transportation Systems of Several Major Cities" could be shortened to "Analysis of Urban Transportation Systems." Note that *Analysis and Comparison* can be replaced by *Analysis*, and *Several Major Cities* can be replaced by *Urban*. However, shortening the title to "Analysis of Transportation Systems" would incorrectly imply inclusion of air, rail, and ship transportation.

Abstract

The abstract is a concise summary of the significant items in a report. Typically, an abstract contains between 200 and 400 words. It should include what has been studied, significant results, and conclusions. Simply stating that transportation systems were analyzed is insufficient. The results of the analysis should also be stated. For example, *In analyzing the transportation systems, it was found that subways are the most efficient means of transporting large numbers of people.* The abstract should also report significant findings. For example, rather than state *Radiation pressure was measured using a torsion balance technique*, write *Using a torsion balance technique, radiation pressure was measured to be* 7.01×10^{-6} *nt/m^2 versus a predicted value of* 7.05×10^{-6} *nt/m^2.* In combination with the title, the abstract should indicate the content of the report. Abstracts of technical papers are often published separately. Therefore the abstract must be able to stand alone without reference to figures, tables, or anything else in the body of the paper.

Introduction

This section introduces the reader to the topic of the report. The introduction contains the objectives of the paper and important background information. Pertinent literature may be surveyed. The introduction usually ends with a very specific statement of purpose.

article be included. A reference may be cited several times using the same number, but it should appear only once in the reference section.

The Latin abbreviation *ibid.* (*ibidem* – in the same place) may be used when an information source is used in subsequent references, provided there are no intervening references cited. In other words, if two or more consecutive references are from the same source, *ibid.* would be used. Note that *ibid.* is a common-enough occurrence in scholarly writing that it is not usually written in italics.

A system of referencing used in many British journals uses the author's name (or authors' names) with year of publication either directly cited or in parentheses. In this case, references are listed alphabetically by author's last name in the reference section.

For example, the citations in the text might be in one of the following formats:

> … and to tension or torsion in the other (TAYLOR and QUINNEY 1931; SCHMIDT 1932)

> … however, it is extremely difficult to check as to whether this is so, as PUGH has recently recognized (1953)

> … as modeled by AVRAMI (1939, 1941)

> … as in CAHN (1956a) and CAHN (1956b)

Note that reference to more than one paper by the same author(s) in the same year is handled by adding letters to their citations.

In the reference section, the citations are listed alphabetically:

AVRAMI, M. (1939), *J. Chem. Phys.*, vol. 7, 1103.
AVRAMI, M. (1940), ibid., vol. 8, 212.
CAHN, J. W. (1956a), *Acta Metall.*, vol. 4, 449.
CAHN, J. W. (1956b), ibid., vol. 4, 572.
PUGH, H. LL. D. (1953), *J. Mech. Phys. Solids*, vol. 1, 284.
SCHMIDT, R. (1932), *Ingenieur-Archiv*, vol. 3, 215.
TAYLOR, G. I., and QUINNEY, H. (1931), *Phil. Trans. Roy. Soc.* A, vol. 230, 323.

Some journals may prefer references by the same author to be listed in one item:

AVRAMI, M. (1939), *J. Chem. Phys.*, vol. 7, 1103; (1940) ibid., vol. 8, 212.
CAHN, J. W. (1956a), *Acta Metall.*, vol. 4, 449; (1956b) ibid., vol. 4, 572.
PUGH, H. LL. D. (1953), *J. Mech. Phys. Solids*, vol. 1, 284.

Appendices

Details of calculations, derivations of equations, or documentation of computer codes that are not essential (but are still valuable) to the presentation of the report can be placed in appendices. Often, detailed

As in technical papers, each table and figure should be numbered, referenced in the text, and appear in the order in which it is referenced. Tables and figures should appear at the end of short letters, since their inclusion in the text may be disruptive to the reader. Footnotes may be used for referencing previous work. Acknowledgments are not required.

Sometimes during the course of an investigation a new discovery unrelated to the main objective is made. This new information can be reported in an appendix. Unrelated significant findings should be reported in a separate letter; otherwise they may be obscured by the requested report.

Summary

The technical letter should end with a summary statement to provide closure. This summary may include unrelated discoveries made during the investigation.

Example Letter 1

This technical letter responds to a request for failure analysis of CDA 260 (cartridge brass) tubes that showed cracking shortly after they were formed into a 180° bend for a heat exchanger.

Date: April 01, 2007
To: Mary C. Haroney, Plant
            Metallurgist
            Brass Tube, Inc.
From: John H. Holliday, Materials
            Engineer
            Technical Center
Subject: Failure analysis of cracked
            CDA 260 tube assemblies

Four cracked CDA 260 copper tubes
were submitted for failure analysis.
Cracks were observed on the inside
radius of the 180° bends of the cool-
ing tube assembly. Stress corrosion cracking was determined to be
the failure mode. A combination of
residual stress and exposure to an
ammonia-based chemical is believed
to be responsible. A stress-relieving
heat treatment of one hour at 260°C
(500°F) is recommended.

Four exemplar tubes were received
for failure analysis from production
lot #1257. The tubes were manufac-
tured from the Copper Development
Association (CDA) alloy 260, which
is a copper alloy containing 30 weight

percent zinc. Circumferential cracks were observed on the inner radius of the bent tubes. Cracked tubes were abrasively cut to expose the fracture surfaces and then examined using a scanning electron microscope. These cracks originate at the surface and extend approximately 400 μm into the tube wall (see Figure 1). Figure 2

Figure 1. A secondary electron image of the exposed crack showing an intergranular fracture, originating at the inside radius of the tube bend. The intergranular cracks extend 400 μm into the tube wall. Figure 2 is located by the box.

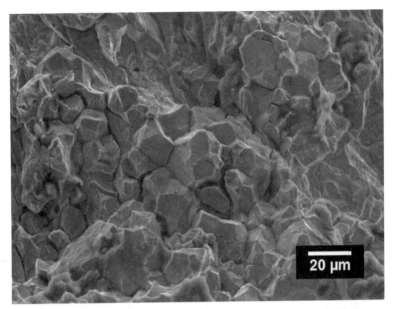

Figure 2. A secondary electron image of the fracture surface that shows an intergranular fracture mode. Intergranular fractures are characterized by a ''rock candy'' morphology of individual crystals or grains.

shows an intergranular fracture path that is typical of stress corrosion cracking of CDA 260.

Locations of the cracks relative to the bent tubes indicate a tensile residual stress resulting from elastic spring-back after bending. The combination of residual stress and exposure to ammonia-based chemicals is known

measured to be 3.25 ± 1.63 µm (95% CL)
using a mean linear intercept method.
Procedures and calculations for this
measurement are shown in the attached
appendix.

The measured hydrogen content is
more typical of the as-cast product.
Annealing will typically reduce the
hydrogen to acceptable levels provided the annealing is performed in
a slightly oxidizing environment.
It should be noted that the oxygen
content reported in Table I is 0.01
weight percent, which is low compared
to the more typical range of 0.1 to
0.15 weight percent for the annealed
product.

In summary, the missile fin parts
from production run #42357 should be
accepted provided the customer waives
the limitations on hydrogen content.
If the customer rejects these parts,
they may be salvaged by a second
annealing heat treatment. The Qual-
ity Control Department will immediately inspect the annealing furnaces
to determine if possible changes in
the furnace operation or personnel may

have affected the heat treatment of the missile fins.

Appendix: Determination of the Mean Linear Size of Alpha Plates

The volume fraction of the β-phase was determined to be 0.20 ± 0.04 (95% CL) by a standard point-counting technique using a 7×7 grid. The grid was placed randomly on the microstructure (see Figure A) ten times to obtain a confidence level (95% CL) that was less than 20 percent of the average value. A mean linear size of the α-phase plate width was then determined to be 3.25 ± 1.63 μm (95% CL) by placing a line of length 130 μm on the photograph as shown in Figure A. Thirty-two α-plates were intercepted, and the mean linear size, L_3, was calculated using

$$L_3 = \frac{\left(1 - V_f^\beta\right) L_{\text{tot}}}{N_{\text{tot}}^\alpha} = \frac{(1 - 0.2)\, 130\,\mu m}{32} = 3.25\,\mu m,$$

where V_f^β is the volume fraction of the β-phase, L_{tot} is the length of the line on the photograph at the image used consistently throughout the presentation. Keep text and graphics away from the screen edges to prevent them from being obscured when the screen is not big enough for the projection system.

Colors and Special Effects

Use strong contrasting colors in all the graphics; it is likely that at least one person in the audience is color-blind. Backgrounds that transition from dark to light only serve to reduce the amount of usable area on the slide and, as a result, force the use of smaller type. The number of organizational affiliations that appear on each slide should be minimized; the audience only needs to read the affiliations once. Clip art should be used sparingly and animations avoided unless they demonstrate specific points in the presentation. Bouncing icons and sound effects only serve to distract the audience.

Chapter Summary

Simple presentations force the audience to focus on the speaker. Keep in mind that many of the great orators in history had nothing more than note cards from which to read. The goal of a presentation is to have the audience remember the ideas and invest in the technology. If the presentation is at a convention and a new job is of interest, remember that there are potential employers in the audience. In this case, it would be prudent to have a resumé handy. Chapter 7 describes some basic guidelines for preparing a resumé.

5 Presentation of Technical Data

This chapter covers some basic guidelines for presenting technical data. Tables, schematic drawings, photographs, and graphs quickly convey technical information and experimental results. A well-prepared table or figure immediately describes the significance of the work and provides a useful tool for the reader. Preparing tables and figures should be treated with the same care as writing a resumé. Technical reports are often the basis of patents, and the use of the standard international system of units (SI) is required for foreign and domestic patent applications. Appendix IV gives the international system of prefixes and units. Use of SI units is recommended, but the author should report in units and symbols most appropriate for the subject and the audience. Common uses of the Greek alphabet are provided in

Appendix V. Results can also be reported using multiple units. This chapter also provides guidance for preparing graphs with multiple scales in different units.

Tables

Tables are helpful for presenting and archiving experimental data. Unlike graphs, tables preserve exact numbers for future analysis. Tables should be sequentially numbered (often with Roman numerals) in the order of presentation in the text. Alphanumeric numbering is used in long reports that are broken into sections. Each table should have a simple title at the top. Tables should be incorporated into the body of the text as soon as convenient after they are referred to (they should not be placed in the middle of a paragraph unless the paragraph breaks across pages), or they can be grouped at the end of the report.

Tables make comparing data easy, so they should be constructed to simplify comparison. Numbers to be compared should be in adjacent columns or rows or both. The structure of Table 5.1 follows these guidelines: comparable heat treatments are in adjacent columns and the numbers to be compared are in the same row.

Table 5.1. *Mechanical properties of an aluminum 2219 alloy tested in tension*

	Test results for naturally aged (T4)	2219-T4*	Test results for artificially aged (T6)	2219-T6*
Young's modulus, GPa	71	71–73	72	71–73
Proportional limit, MPa	110		220	
Yield strength at 0.2% offset, MPa	136	185 min.	295	290 min.
Ultimate tensile strength, MPa	317	360 min.	426	415 min.
Percent elongation, 0.5-inch gauge	26	20 min.	16	10 min.

* Properties for 2219 are from Hatch (1984). Percent elongation was reported for a 2 inch gage length.

Figures and Figure Captions

Figures include any visual material – except tables – that may aid the reader. Figures should be numbered in the order in which they appear in the text; that is, the fifth figure referred to in the text should be Figure 5. All figures should be referred to in the text. Each figure should have a caption underneath it (not a title above it) that provides enough descriptive text to allow the figure and caption to stand alone and convey a significant finding. The caption, which may be several sentences long, should tell the reader what to look for in the figure. The number of figures should be limited to those necessary to justify the conclusions of the report.

Schematic Drawings

Schematic drawings should be labeled to indicate important parts. Do not expect the reader to know the features of the drawing. The same applies to features in photographs. Figures 5.1 and 5.2 are schematics of an electrical circuit and material flow in a chemical process.

Schematic drawings can also provide the basis of a theoretical analysis, as shown in Figure 5.3 for hardness testing, or show how a process works. Figure 5.4 shows schematically where a solid mandrel is placed in a tube while the tube is being bent. Note that the figure captions explain each of these drawings.

Equipment Photographs

Photographs of equipment or equipment models should clearly identify the important features. Figure 5.5 shows a conceptual model of a modern recoilless rifle. An alternative to labeling features in a photograph is shown in Figure 5.4, where each part is labeled with a letter and explained in the figure caption.

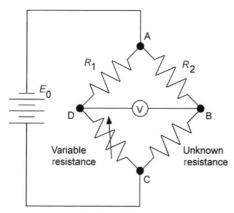

Figure 5.1. A Wheatstone bridge for measuring resistance with an applied voltage of E_0. The resistances R_1 and R_2 are equal. The variable resistance is adjusted until there is no voltage between D and B. The value of the unknown resistance equals the resistance of the variable resistor at that point.

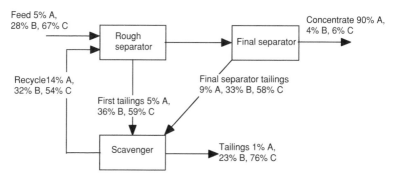

Figure 5.2. Schematic of a separator designed to enrich the concentration of A. Tailings from rough and final separators are recycled.

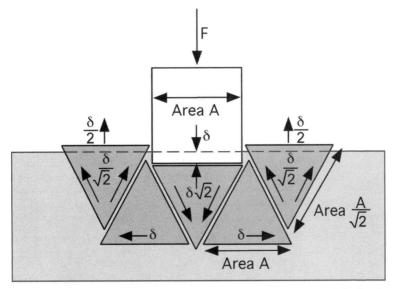

Figure 5.3. A schematic diagram used in the analysis of the hardness indentation. The five triangular portions slip relative to each other performing work under the applied load and displacement δ of the indenter. Figure adapted from Ashby and Jones (1981).

Photomicrographs

A scale bar in the photograph should be used to indicate magnification. Stating the magnification in the caption is not required or beneficial because the publisher may alter the size of the figure or the report may be reproduced at a different magnification. When the figure shows a microstructure, it is standard practice to include in the caption the chemical etchant used to reveal the structure. Each phase

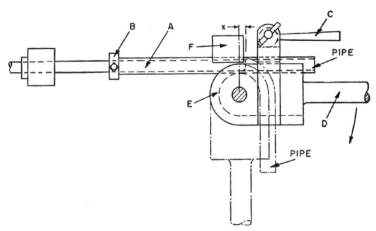

Figure 5.4. Drawing of a mandrel through a tube while it is being bent preserves the circular cross section. In the upper sketch, the tube to be bent is pushed over a mandrel (A) and against a stop (B), which locates the bend. The tube is clamped by a lever (C) and pulled by a lever (D), causing the form (E) to rotate (Schubert 1953).

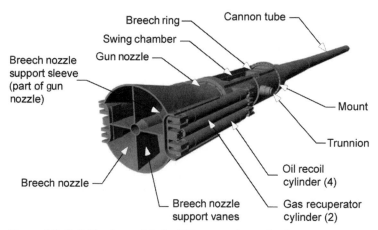

Figure 5.5. Solid-body model of a 105-mm sonic rarefaction wave gun or recoilless rifle (Kevin Miner, Benet Laboratories, September 15, 2006).

should be identified in both the figure and the caption (see Figure 5.6). One should not assume that the reader can identify the phases. However, some microstructures, such as the titanium microstructures shown in the second example letter of Chapter 3, do not lend themselves to labeling; such microstructures should be explained in the caption. When reports are reproduced, the significant features used to identify the phases may become obscured in the copies. Thus, clear labeling and descriptive captions are essential to the integrity of the report.

Graphing

Graphs are a good way of presenting data so the reader can see trends. Although most graphs are now produced using personal computers, there are some common pitfalls to avoid. Avoid the use of background colors or shading as they can obscure the data. Using color for data points should also be avoided. Most publications are in black and white, and grayscale renderings of shaded graphs or lightly colored data points are difficult to see. Do not include a title for the graph; that information should be in the caption. Enclosing the graph in a box wastes space. As a general rule, the graph should be legible after it has been converted to grayscale and reduced to

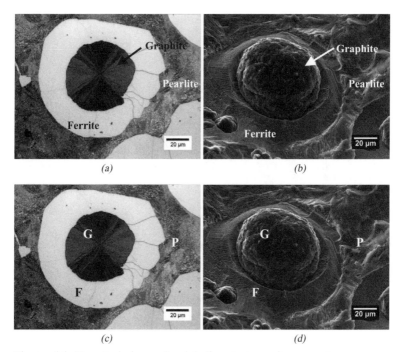

(a)   *(b)*

(c)   *(d)*

Figure 5.6. Two techniques for labeling features in an optical image of a nodular iron microstructure. In (a) and (b) the microstructure is labeled directly and the reader need not read the caption to interpret the image. However, large portions of the image can be obstructed using this technique. Symbols (F-ferrite; P-pearlite; G-graphite) are used in (c) and (d) to label the microstructure. The images are of a ductile iron where the graphite forms nodules. Polarized light was used to show the radial growth of the graphite nodule. A 2 percent nital etchant (2 volume percent nitric acid in ethanol) was used to reveal the ferrite and pearlite microstructure in the optical images (a) and (c). The nodule is encapsulated in ferrite (F) and the combination is often referred to as the bull's-eye microstructure. The ferrite grains are attacked along crystallographic planes when deep etched (5 volume percent bromine in methanol), as revealed by the secondary electron images shown in (b) and (d).

75 mm by 75 mm size. The following sections provide some basic guidelines on the science of graphing.

Ordinate vs. Abscissa

Normally, one plots the dependent variable on the ordinate (y-axis) and the independent variable on the abscissa (x-axis). For example, suppose the current through a rectifier is studied as a function of voltage; the current is the dependent variable and should be plotted on the ordinate. Another example is gasoline consumption by an airplane with respect to speed. In this case, gallons per mile is the dependent variable and velocity is the independent variable. Sometimes there is confusion on this point in stress-strain curves. Conventionally, stress is represented as the dependent variable (y-axis) and strain as the independent variable. This corresponds to the way in which most tension tests are made. The testing machine forces an elongation, and we measure the resulting force, which depends on the test bar. If one were to make a tension test by dead weight loading (e.g., adding a known weight of sand to a bucket hung from the test bar), the load (stress) would be the independent variable. However, in this case, plotting stress on the ordinate is still preferred because this is the conventional way of representing stress-strain curves, and therefore it would be most easily interpreted by readers.

There are other occasions when this convention is not followed. For example, fatigue data are usually presented in the form of an *S-N* plot with the stress, *S*, as the ordinate and cycles to failure, *N*, as the abscissa. Clearly, stress is the independent variable and cycles to failure is the dependent variable.

There are also cases in which there is no clear-cut choice about which variable is independent and which is dependent. An example is a graphical correlation between the weights and wingspans of airplanes. Neither measure is more independent than the other. The author's judgment should be used in such cases.

Choosing Scales

The first step in making a graph is to select the scales. In general, the scales should be selected so that the data cover a reasonably large fraction of the graph. This is illustrated in Figure 5.7.

The divisions on the graph should represent multiples of 1, 2, or 5×10^n units (but not 3, 12, etc.), as shown in Figure 5.8.

If the slope of a plotted line is important for interpretation, the scales should be adjusted so that

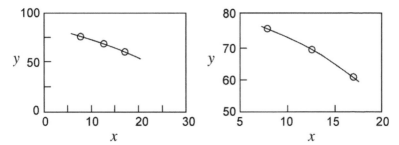

Figure 5.7. Choosing scales so the data occupy a large portion of
the plot is important. The plot on the left could be expanded with
an ordinate range of $50 \leq y \leq 80$ instead of $0 \leq y \leq 100$ and an
abscissa range of $5 \leq x \leq 20$ instead of $0 \leq x \leq 30$, as shown on
the right.

the slope is in the range of 30 to 60 degrees. It is very
difficult for the reader to check a slope that is nearly
horizontal or nearly vertical.

Whether the origin should be shown depends
on several factors. One is the nature of the quantity
being plotted. For example, if temperatures are being
plotted in Fahrenheit, $0°$ has no special significance
so there is no compelling reason to start the scale at

okay	0	5	10	15	20
or	0	10	20	30	40
or	0	0.02	0.04	0.06	0.08
but not	0	3	6	9	12
or	0	13	26	39	52

Figure 5.8. Examples of appropriate and inappropriate scale
divisions.

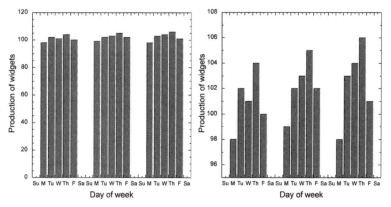

Figure 5.9. Inclusion of zero on a scale can obscure the importance of trends, as shown in the figure on the left. The figure on the right makes the differences appear greater and clearly shows that fewer widgets are produced on Monday than on any other day, that production steadily increases during the week, and that production decreases on Friday.

zero. The same is true of sound in decibels because there is no special significance to zero.

A second factor to be considered is whether one can show the origin and still have large enough divisions to show important variations. Consider a plot of how daily widget production varies through a three-week period (see Figure 5.9) when the daily production varies by less than 8 percent. Inclusion of zero on the y-axis makes the variation in widget production more difficult to see.

A third consideration is whether the data are being tested against or compared with a theory in which the origin has a special significance. Suppose the thickness of a chemical reaction product, L, has been measured as a function of time, t, and one wishes to compare the data with a theory that predicts that L is proportional to $\sqrt{t}$ (this means that $L \to 0$ as $t \to 0$). In this case, it would be appropriate to plot L vs. $\sqrt{t}$ on scales that do include the origin so one can see whether a straight line through the data extrapolates to the origin.

Labeling Scales

It is not necessary to indicate the scale level at every line. If too many lines are labeled, the graph will look cluttered; if too few are labeled, the scale will be difficult for the reader to interpret. A reasonable compromise is to indicate the scale level every 2 to 5 places, as shown in Figure 5.10.

It should be easy to tell which numbers on a scale refer to a major division. This can be a problem when the numbers are very large or very small, as indicated in Figure 5.11. In this case it is better to either plot 0.0005 as 0.5 on a $1000/T$ scale or to label it as 5×10^{-4}. Avoid ambiguity when labeling scales. For example, $1000/T$ indicates that the reciprocal of the temperature has been multiplied by 1000. This can also be written as $1/T \times 10^3$, but $1/T \, (10^3)$ is ambiguous and should not be used.

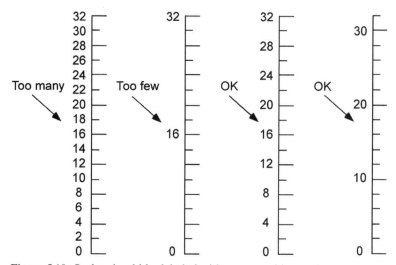

Figure 5.10. Scales should be labeled with a reasonable number of divisions. Too many labels make the scale difficult to read, whereas too few requires the reader to determine the values.

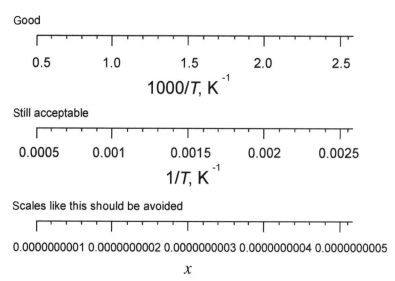

Figure 5.11. Examples of the appropriate use of numbers on axes to avoid overcrowding and a scale that is too crowded and difficult to read.

Use a font size of at least 14 points for the numbers in the scale and at least 16 points to label the axes. These guidelines will help retain readability when the graph is reduced for publication. A smaller font may be justified to avoid overcrowding. In Figure 5.9, 12-point type was used to avoid overcrowding the x-axis scale. However, expanding the x-axis and stacking the figures vertically rather than horizontally would avoid the overcrowding and permit the use of larger type.

Multiple Units on a Single Scale

Sometimes multiple units are used to describe the data in a graph. For example, stress data are often represented in both SI (MPa) and English (psi) units. If the left ordinate is labeled with SI units, the right ordinate can be in English units, or vice versa. The scale on the left ordinate can be indicated with tick marks at intervals of 1, 2, or 5×10^n. Intervals on the right ordinate should not coincide with intervals on the left ordinate; if they do, the numbers will not be simple (see Figure 5.12).

The same concept applies to labeling the axes to show both weight percent and atomic percent, both engineering strain and true strain, and the like. Remember, the axes must be labeled clearly to indicate both the variable being plotted and its units.

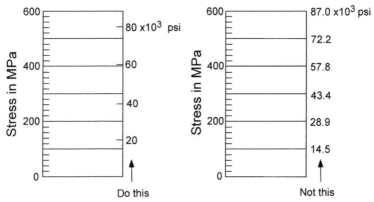

Figure 5.12. Appropriate use of more than one set of units.

Points

Experimental points should be plotted and appear large enough that they will be clearly visible even if the figure is reduced for publication. Calculated points are not generally indicated when a curve is theoretical. Error limits on the points may be included in the plot; the caption should indicate the level of precision or the amount of uncertainty that the limits represent.

Multiple Plots

Frequently, it is advantageous to plot more than one curve on the same graph. This saves paper and publication space (space costs money!). More importantly, it allows the reader to compare the curves.

When two or more curves are plotted on the same axes, it is essential that the reader be able to easily tell which curve is which and which points belong to each curve. If the curves are well separated and all of the points lie close to the curves drawn through them, it is sufficient to simply label each curve. However, it may be necessary to use different symbols for points on different curves ($\triangle$, $\blacksquare$, $\blacksquare$, $\circ$, $\square$, $\diamond$, etc.) or different types of lines for different curves (-----, ——,......,-·-·- -·-·-, etc.). In these cases, it is necessary to include a legend explaining the points and lines. An example of using different plot symbols and lines is shown in Figure 5.13 for the fatigue life of test specimens machined from titanium sheet by water-jet cutting and specimens that were subsequently polished to remove the damage created by water-jet cutting.

Drawing Curves Through Experimental Points

Should curves be drawn through every point or should they be smoothed to follow the general trend? Or should the data be approximated by a straight line? The answers depend on the circumstances. If there is reason to believe that the data are of sufficient accuracy that each hump and dip is real, then the curve should be drawn through all the points.

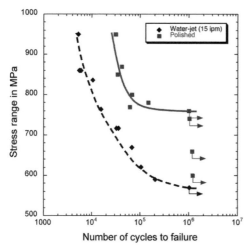

Figure 5.13. An example of two data sets being plotted in the same graph. The legend position should be chosen so that it does not obscure the data or interfere with reading the graph.

Consider Figure 5.14(a). Perhaps y is the temperature of a furnace and x is the time after it is turned on; the hills and valleys correspond to the cycling of the controller.

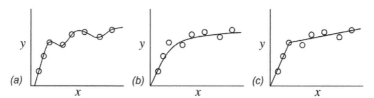

Figure 5.14. Three ways of drawing curves through experimental points: (a) curve drawn through all points, (b) smooth curve approximating points, and (c) data represented by two straight lines.

On the other hand, the same points can be represented by a smooth curve as shown in Figure 5.14(b), unless the deviation of such a curve from the points is greater than the possible error of the measurement or there is some compelling reason to believe that the phenomenon is cyclic. For example, y may represent the thickness of a growing film and x time. Finally, theory may play a role in how the data are best presented. Perhaps theory suggests that the behavior can best be represented by two straight lines, as in Figure 5.14(c).

Finding Slopes of Straight Lines

A straight line can be represented by the equation $y = ax + b$. The slope, a, can be found from two points, (x_1, y_1) and (x_2, y_2) as $a = (y_2 - y_1)/(x_2 - x_1)$. Two well-separated points should be used to avoid errors caused by experimental scatter (see Figure 5.15).

Grid Lines

The person making the graph should decide whether grid lines would be helpful to the reader. Grid lines may be useful if a reader needs to read specific values from the graph. They should be omitted if they obscure data or trend lines. A good compromise is to

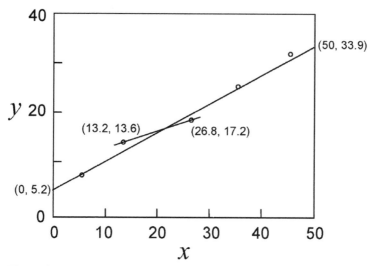

Figure 5.15. Determining the slope from points too close to one another can lead to great error. Here the true slope is $(33.9 - 5.2)/(50 - 0) = 0.57$, not $(17.2 - 13.6)/(26.8 - 13.2) = 0.26$.

use tick marks on the right and top scales as well as on the ordinate and abscissa.

Logarithmic Scales

Logarithmic scales are often labeled only at intervals differing by factors of 10 with no intermediate grid lines. If x is plotted on a logarithmic scale, the distance between two values x_1 and x_2 depends on the ratio of x_2/x_1. The distance on the paper between 1 and 2 is the same as the distances between 2 and 4 and between 5 and 10. In reading values between

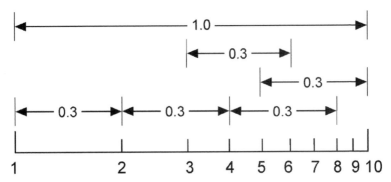

Figure 5.16. Reading a logarithmic scale. Note that the paper distance between two points that differ by a factor of 2 is close to 3/10 the length of the decade. A single decade is shown in the figure.

1 and 10, keep in mind that 2 is at a point about 0.3 times the distance between 1 and 10, so 5 is represented by a point about 0.7 of the distance between 1 and 10 (see Figure 5.16).

There are several reasons for using logarithmic scales. Often, the quantity being plotted varies by factors of 10, 100, 1000, or more over the range of interest and we want to be able to distinguish 8 from 10 as much as 800 from 1000. Other times, there are theoretical reasons for using logarithmic scales. In these cases there are two options: one is to plot the logarithm of the quantity directly on a Cartesian scale. In this case the scale should be labeled accordingly. The disadvantage of this approach is that it is difficult for the reader to discern the real value of the quantity. The other option is to use a logarithmic

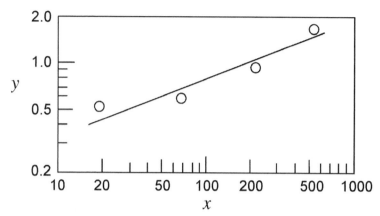

Figure 5.17. Labeling axes on a logarithmic scale. For the example shown, the additional labels on the y-axis are important, since only a partial decade appears on the y-axis.

scale and label it with the quantity directly. Often it is necessary to label the scales at a reasonable number of intervals (e.g. 0.02, 0.05, 0.10, 0.20, 0.50, 1.0, 2.0), as indicated in Figure 5.17.

Finding the Slope on a Log-Log Plot

If $y = ax^b$, then $\ln(y) = \ln(a) + \ln(x)$ and a plot of $\ln(y)$ vs. $\ln(x)$ on Cartesian coordinates or a plot of y vs. x on log-log paper will have a slope equal to b. The simplest way to find a slope is to take two well-separated points and realize

$$b = \frac{\ln(y_2) - \ln(y_1)}{\ln(x_2) - \ln(x_1)} = \frac{\ln(y_2/y_1)}{\ln(x_2/x_1)}.$$

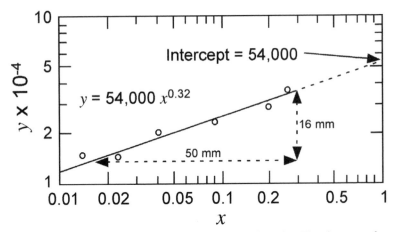

Figure 5.18. Finding the slope on a log-log plot. The slope may be determined by direct measurement using a scale where the rise and run are measured on the paper. In this example, the slope is equal to 16 mm/50 mm or 0.32.

Alternatively, the slope can be found by simply using a ruler to measure the x-distance and the y-distance and correcting them for the distance on the plot of a decade:

$$b = \frac{\Delta y \text{ in mm}/y \text{ decade in mm}}{\Delta x \text{ in mm}/x \text{ decade in mm}}.$$

Do not use the numbers on the scales of the log-log paper to determine the slope directly.

Note that if $y = ax^b$, a equals the value of y where $x = 1$ (see Figure 5.18).

Finding the Slope on a Semi-Log Plot

Arrhenius rate equations are often encountered in physical chemistry; the activation energy, Q, can be determined from a semi-log plot. The temperature dependence of diffusivity is a typical example of an Arrhenius-type relationship:

$$D = D_0 \exp \frac{-Q}{RT}.$$

A semi-log plot for the diffusivity of silver is shown in Figure 5.19. In this example the slope is equal to $-Q/R$, where R is the universal gas constant. Unlike the log-log determination for the slope, the actual values used to calculate the slope in a semi-log plot must be determined from the graph. For the exponential relationship the following formula can be used to calculate the slope:

$$slope = \frac{\ln(y_2/y_1)}{x_2 - x_1}.$$

In the example shown in Figure 5.19, the slope would be calculated as

$$slope = \frac{-Q}{R} = \frac{\ln\left(10^{-9}/4 \times 10^{-11}\right)}{(0.904 - 1.065)} \times 1000$$

$$= -19,990 \text{ K},$$

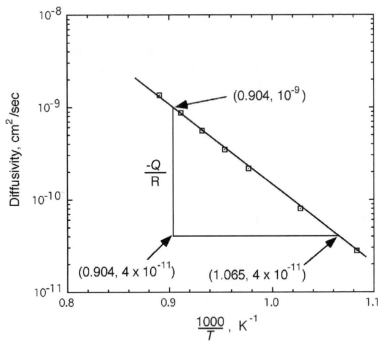

Figure 5.19. A semi-log plot showing the diffusivity of Ag in Ag as a function of the reciprocal temperature. The slope of the trend line is equal to $-Q/R$, where Q is an activation energy and R is the universal gas constant.

and the activation energy, Q, is determined as

$$Q = -R \times slope = -8.314 \, \frac{J}{mol \cdot K} \times -19{,}990 \, K$$

$$= 166{,}200 \, \frac{J}{mol}.$$

Figures Generated by Computer Screen Prints

Computationally intensive computer programs are becoming an increasing part of the science and engineering professions. Finite element analysis for stress calculations and computational fluid dynamics for fluid flow and heat transfer are the most common examples. Figures showing calculated results are often generated as screen prints, but what appears to be readable on the computer screen is seldom legible when reduced for publication. An example is shown in Figure 5.20 for the calculated temperature profile of two aluminum bars that are 17.8 cm (7 inches) in diameter that are being heated by hot gas flowing from left to right. Figure 5.20(a) shows a typical screen print where the temperature scale is in scientific notation and the units of temperature and time are not explicitly defined. The same calculation is redrawn in Figure 5.20(b) to be more reader-friendly. The only information missing from Figure 5.20(b) is the time lapse for the calculation; this information should be included in the caption.

Chapter Summary

This chapter dealt with the visual presentation of data and best practices for graphing. It is often desirable to show a relationship between measured quantities and

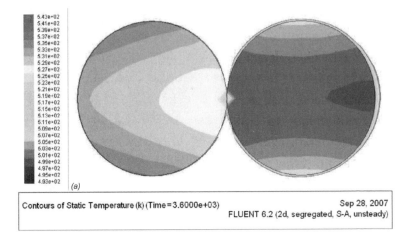

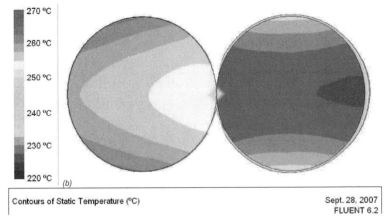

Contours of Static Temperature (ºC)   Sept. 28, 2007
                                                                    FLUENT 6.2

Figure 5.20. Screen prints generated from the computational fluid dynamics program FLUENT 6.2. The results show the temperature profile of two 17.8-cm-diameter aluminum cylinders after one hour of heating by a hot gas flowing from left to right. (a) An example of a screen print that is difficult to read because of the temperature scale format. In addition, the screen image does not specify the units of time, and the units of temperature should be capitalized, that is, K rather than k. (b) The same figure reformatted to make the temperature scale more readable.

Table 6.2. *The t values for calculating the 95 percent confidence levels*

$N-1$	t	$N-1$	t
1	12.706	18	2.101
2	4.303	19	2.093
3	3.182	20	2.086
4	2.776	21	2.080
5	2.571	22	2.074
6	2.447	23	2.069
7	2.365	24	2.064
8	2.306	25	2.060
9	2.262	26	2.056
10	2.228	27	2.052
11	2.201	28	2.048
12	2.179	29	2.045
13	2.160	30	2.042
14	2.145	40	2.021
15	2.131	60	2.000
16	2.120	120	1.980
17	2.110	∞	1.960

Source: Dieter 1991.

the mean at a 95 percent confidence level, $\Delta\bar{x}$, is then given by

$$\Delta\bar{x}(95\%\,CL) = \frac{ts}{\sqrt{N-1}}. \qquad (7)$$

Test for the Normal Distribution

Positive and negative deviations will occur with equal probability only if the distribution is normal. A convenient way of determining whether a set of data

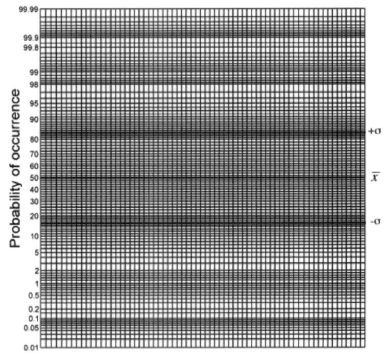

Figure 6.2. Normal probability paper. The standard deviation is
determined as the difference between the arithmetic mean, $\Delta\bar{x}$,
and the value at $+\sigma$ or $-\sigma$.

exhibits a normal distribution is to plot the cumula-
tive frequency on normal probability paper, which is
shown in Figure 6.2. The ordinate of the graph paper
is distorted in such a manner as to produce a straight-
line plot when the data have a normal distribution. To
plot N measurements, the data are sorted in increasing order and given a rank, i, starting with the number one. An accumulative probability, or probability

the lowest measurement is removed. However, it is not appropriate to arbitrarily exclude measurements that do not meet expectations. Chauvenet's criterion (1989) provides a means to test the data and determine whether a particular measurement can be removed from a data set. It should be emphasized that this procedure allows only one measurement to be removed.

To apply Chauvenet's criterion, the arithmetic mean and the standard deviation are calculated for the data set in the usual manner. For small data sets the standard deviation (σ) can be approximated by the sample standard deviation (s). In addition, the ratio of the deviation, d_i, to the standard deviation, σ, is calculated for each measurement using Equation (10); these results are also shown in Table 6.3 for the fracture stress of the carbon-fiber composite:

$$\frac{d_i}{\sigma} = \frac{|x_i - \overline{x}|}{\sigma}. \tag{10}$$

Chauvenet's criterion requires that the ratio calculated using Equation (10) must exceed a specified value before the measurement can be excluded; this value depends on the number of tests, N (Table 6.4). Chauvenet's criterion assumes a normal distribution. According to Table 6.4, the maximum deviation for the group of 19 measurements is between 2.13 and 2.33. The largest deviation of the data in Table 6.3 is

Table 6.4. *Chauvenet's criterion for rejecting a measurement*

Number of measurements, N	Ratio of maximum deviation to standard deviation, d_{max}/σ
3	1.38
4	1.54
5	1.65
6	1.73
7	1.80
10	1.96
15	2.13
25	2.33
50	2.57
100	2.81
300	3.14
500	3.29
1,000	3.48

1.98, so all of the data must be included in the statistical analysis. If Chauvenet's criterion is met, then the arithmetic mean and standard deviation are recalculated after removing the dubious measurement. The value of N must also be reduced by one.

Weibull Statistics

Most physical properties exhibit a lower bound in the probability distribution, which the normal distribution fails to accurately describe. The Weibull distribution was originally proposed for describing fatigue life, but it has been used to model many different engineering properties, such as brittle fracture of ceramics and the life of electronic components. The probability density, $p(x)$, for the Weibull distribution is given by

$$p(x) = \frac{m}{\theta} \left(\frac{x}{\theta}\right)^{m-1} \exp\left[-\left(\frac{x}{\theta}\right)^{m}\right]. \tag{11}$$

The shape of the distribution curve is controlled by the value of m and is referred to as the Weibull modulus. Example distributions, with varying values of m, are shown in Figure 6.4. The population distribution narrows rapidly as the value of m increases, and measurements with a high Weibull modulus are thought of as more reliable because there is less scatter in the data. The scaling parameter θ is called the characteristic value; at $x = \theta$ the population is divided into 63.2 percent below and 36.8 percent above θ for all values of m.

Calculation of a mean, Equation (12), and of a variance, Equation (13), for a Weibull distribution is not straightforward; these calculations involve the standard gamma function, Γ. However, the main reason for using Weibull statistics is not to report means or variances, but rather to report the probability of an event occurring.

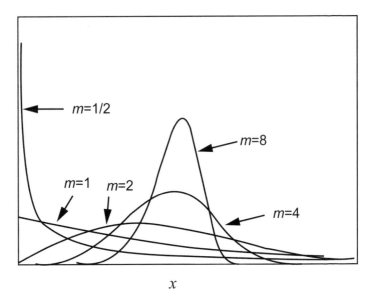

Figure 6.4. A schematic plot showing the Weibull distribution function with different values of m. In this plot $\theta = 1$ and $x_0 = 0$.

$$\bar{x} = \theta\Gamma\left(1 + \frac{1}{m}\right). \qquad (12)$$

$$\sigma_p^2 = \theta^2\left[\Gamma\left(1 + \frac{2}{m}\right) - \Gamma^2\left(1 + \frac{1}{m}\right)\right]. \qquad (13)$$

The probability of occurrence, $P(x)$, for the Weibull distribution is easily plotted with any graphics software package and the important parameters obtained graphically or by linear regression analysis. To incorporate a lower bound to the population, a third parameter, x_0, may be introduced. The three-parameter equation for $P(x)$ is given by

$$P(x) = 1 - \exp\left[-\left(\frac{x - x_0}{\theta - x_0}\right)^m\right]. \tag{14}$$

The probability of seeing a value less than x_0 is zero. Setting x_0 equal to zero in Equation (14) produces a standard two-parameter Weibull equation with a probability distribution characterized by Equation (11).

To produce a straight-line plot, Equation (14) is rewritten as

$$\log\left[\ln\left(\frac{1}{1 - P(x)}\right)\right] = m\log(x - x_0) - m\log(\theta - x_0)$$

or

$$\ln\left[\ln\left(\frac{1}{1 - P(x)}\right)\right] = m\ln(x - x_0) - m\ln(\theta - x_0).$$

Values of $P(x)$ are obtained in the same manner as with normal probability paper. The data are first sorted in ascending order and ranked. $P(x)$ is then calculated using Equation (8) or (9), with the exception that a fractional number is used rather than a percentage. The data are then plotted as

$$\ln\left(\frac{1}{1 - P(x)}\right) \text{ vs.}(x - x_0)$$

on log-log axes or on linear scales as

$$\log\left[\ln\left(\frac{1}{1 - P(x)}\right)\right] \text{ vs. } \log(x - x_0)$$

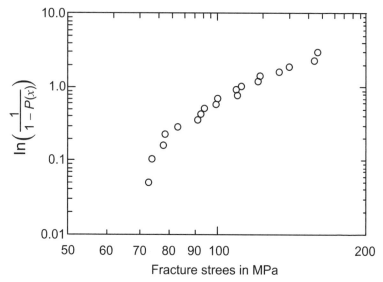

Figure 6.5. A two-parameter Weibull plot of fracture stress for the carbon-fiber-composite data shown in Table 6.3. A first approximation of x_0 for the three-parameter Weibull plot can be found by extrapolating an imaginary curve through the data and down to the abscissa. A value of 70 MPa is found using this method.

or

$$\ln\left[\ln\left(\frac{1}{1-P(x)}\right)\right] \text{ vs. } \ln(x-x_0).$$

A two-parameter plot is shown in Figure 6.5 for the carbon-fiber-composite data in Table 6.3. The negative curvature indicates a value greater than zero for x_0. A first approximation of $x_0 = 70$ MPa is found by extrapolating an imaginary curve through the data and down to the abscissa. The best value of x_0 is found

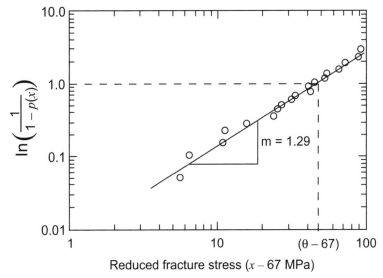

Figure 6.6. A three-parameter Weibull plot of fracture stress for the carbon-fiber-composite data shown in Table 6.3. A value of $x_0 = 67$ MPa was found to produce the best straight line.

by adjusting x_0 and observing the change in curvature. If the data show positive curvature, then the x_0 value is too high. The best straight line for the carbon-fiber-composite data was obtained by setting x_0 equal to 67 MPa; see Figure 6.6. Values for m and θ can then be obtained graphically, as shown in Chapter 5, or more directly using a linear fitting routine when the data are plotted as shown in Figure 6.7.

Extracting engineering information directly from the graph is a little easier if log-log scales are used as shown in Figure 6.6, but Figure 6.7 has the

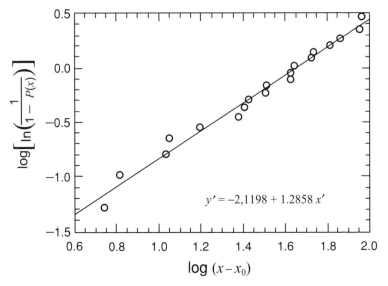

Figure 6.7. An example of the same three-parameter plot as in Figure 6.6, but using a linear scale on the ordinate and abscissa. A linear fitting routine can now be used to determine the equation of the best-fit line.

advantage of yielding the best-fit line from which values of m and θ may be calculated from the following relations:

$$y' = \log\left[\ln\left(\frac{1}{1 - P(x)}\right)\right]$$

$$1.2858x' = m\log(x - x_0)$$

$$-2.1198 = m\log(\theta - x_0).$$

where $m = 1.2858$.

Probability of Failure Calculations

The probability of occurrence, $P(x)$, may also be thought of as the probability of failure when x represents a failure stress or the number of cycles to failure. Weibull analysis then provides information about the probability of failure that can be used in design. For the example of the carbon-fiber composites, a safe loading limit might be specified as 67 MPa because the probability of failure at this stress is zero. If the composite is used in a non–life-threatening application, then perhaps a failure rate of one out of a thousand is acceptable. Using the best-fit equation from Figure 6.7, an applied stress of 67.2 MPa would fail one out of a thousand $(P(x) = 0.001)$ carbon-fiber composites.

Example calculations:

$$\log\left[\ln\left(\frac{1}{1 - 0.001}\right)\right] = -2.99978$$

$$-2.99978 = 1.2858\log(x - 67\text{ MPa}) - 2.1198$$

$$x = 67.2\text{ MPa.}$$

Weibull analysis might also be used to predict the number of hours that a circuit can operate with only 0.0001 percent chance of failure. Instead of testing a million circuits, one can test 100 circuits and get reasonable values of m and θ to solve for x with $P(x) = 10^{-6}$.

Uncertainty Analysis

It is sometimes necessary to transform experimental data and the corresponding uncertainty by a mathematical operation to obtain a desired engineering result. A typical example would be the measurement of a stress. During a tensile test the load, F, rather than the stress, S, is actually measured and must be converted to a stress by dividing by the cross-sectional area, πr^2. Uncertainty in the stress value is introduced as a result of errors in measuring the sample radius, r, and errors in measurement of the load. Converting these errors to an uncertainty in the stress can be accomplished in a number of ways. A commonsense approach is to combine all of the errors in the most detrimental way to determine the minimum and maximum values that might be obtained. The following is an example of this approach for calculating the stress:

$$\frac{\overline{F} - \Delta F}{\pi (\overline{r} + \Delta r)^2} \leq S \leq \frac{\overline{F} + \Delta F}{\pi (\overline{r} - \Delta r)^2}.$$

A more precise method for calculating uncertainties was developed by Kline and McClintock (1953). To illustrate how this method was formulated, consider the relationship between y and x when they are related by the equation $y = \ln x$; see Figure 6.8.

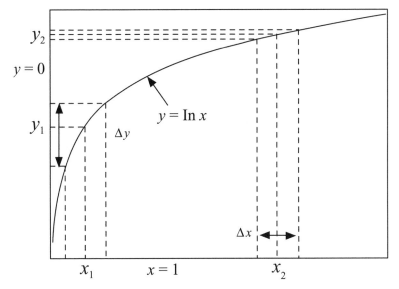

Figure 6.8. A schematic drawing that illustrates how the uncertainty in the x variable may be converted to an uncertainty in the y variable via a mathematical transformation. In this case y is related to x through the equation $y = \ln x$.

The Δy uncertainty will depend on the value of x, through the calculation of the slope at x, and Δx. For this particular example, it should be noted that small values of x produce large Δy uncertainties:

$$y = \ln x$$
$$\frac{dy}{dx} = \frac{1}{x} = \frac{\Delta y}{\Delta x}$$
$$\Delta y = \frac{\Delta x}{x}.$$

This method can be used to convert uncertainties in measurements when a mathematical transformation of the data is necessary to obtain a straight-line plot, for example, plotting $\ln x$ rather than x on a log scale.

When multiple variables $(x, y, z, \dots)$ are used to calculate a quantity, w, and each has an uncertainty $(\Delta x, \Delta y, \Delta z, \dots)$ associated with it, then the following general equation is used:

$$w = f(x, y, z, \dots)$$

$$\Delta w = \sqrt{\left(\frac{\partial f}{\partial x}\Delta x\right)^2 + \left(\frac{\partial f}{\partial y}\Delta y\right)^2 + \left(\frac{\partial f}{\partial z}\Delta z\right)^2 + \dots} \tag{15}$$

The resultant uncertainty will have the same confidence level as the uncertainties used in the calculation, provided they are all the same. Thus, if all the uncertainties are given at the 95 percent confidence level, then the result will also be at a 95 percent confidence level.

Example of Transforming Uncertainties

The hardness of recrystallized cartridge brass is dependent on grain diameter as described by

$$H = H_0 + \frac{k}{\sqrt{L_3}}, \tag{16}$$

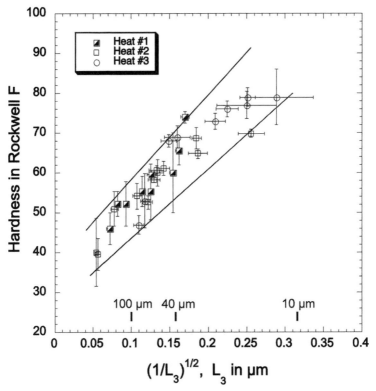

Figure 6.9. Hardness and grain diameter data for recrystallized cartridge brass. Data were collected from three consecutive productions, or heats, of cartridge brass. Uncertainties represent a 95 percent confidence level for the ordinate and abscissa values. A linear relationship is expected between hardness and the grain diameter based on Equation (16).

where L_3 is the mean linear intercept of the grain diameter, H_0 and k are materials constants, and H is the hardness.

A plot of H vs. $1/\sqrt{L_3}$ should produce a straight line (see Figure 6.9), and the uncertainty in L_3 would be transformed to an uncertainty (Δx) on the abscissa as follows:

$$x = \frac{1}{\sqrt{L_3}}$$

$$\Delta x = \sqrt{\left(\frac{dx}{dL_3}\Delta L_3\right)^2} = \frac{1}{2}(L_3)^{-\frac{3}{2}}\Delta L_3,$$

where ΔL_3 is the uncertainty in the measured value of the grain diameter of L_3. Figure 6.9 is a compilation of three consecutive productions, or heats, of the cartridge brass, and the data follow an approximate linear trend as shown by the upper and lower bounds.

Chapter Summary

Uncertainty analysis and reporting confidence levels can add credibility to a technical report; that is why this chapter is included in this book. Although this chapter provides basic guidance in using statistical analysis, it is not a complete treatment. However, the information provided should be sufficient to treat experimental results obtained in most undergraduate science and engineering laboratory courses.

7 Resumé Writing

The purpose of a resumé is to obtain a job. Only a small fraction of resumés result in job interviews. The reader of resumés spends an average of 30 seconds on each one. To be successful, a resumé should be short, with the important information listed first. It should be well organized and neat. Often a resumé is tailored to a specific job, which would require rewriting the resumé for each new position.

Organization

The first step is to gather the pertinent information and organize it. Then this information should be divided into headings such as *Personal Information*, *Work Experience*, *Education*, *Skills*, *Honors*, and *Activities*.

Personal Information

The person's name, without titles, should appear at the top of the page in a larger font than the rest of the document; use this larger size type for the headings as well. Next list home address, phone numbers, email address, and fax number (if applicable). Citizenship may be listed, but this is not required. Personal information such as age, sex, and general health need not be listed either.

Work Experience

Experience includes full-time and part-time jobs, internships, academic research positions, and volunteer work. List the employer, with months and years worked, position title, and responsibilities. For example, *Sam's Café, September 2005 to June 2006, waiter* or *ABC Chemicals, June to August 2005, summer intern, preparing special orders.* Jobs should be listed in reverse chronological order, with the most recent first. Inventions and publications may be listed under Experience or separately under *Publications* or *Inventions* at the end of the resumé.

Education

Education listed on the resumé should include only college-level studies. The exception is if one is only in the first year of college and applying for a summer internship; then high school information may be included. Degrees, with month and year obtained or expected, should be listed, along with the name of the school, major (and minor if any), and grade-point average. Sometimes listing of important courses is appropriate.

Skills

Skills include facility in computer languages, foreign languages, teaching, communication, leadership, and teamwork. List the most important skills first.

Honors

Honors include scholarships, academic awards, and recognition of community service or athletic achievement. These are also listed in reverse chronological order.

Activities

Under Activities, list any student, professional, or community organizations and the various offices (e.g., president, treasurer, secretary) held in these organizations. Listing of extracurricular activities and hobbies is optional.

Wording

Certain words help create a favorable impression. Among these are:

ability	achieved	built	conceived	controlled
demonstrated	developed	devised	directed	enhanced
exhibited	expanded	generated	helped	imagination
improved	incorporated	installed	led	managed
motivated	organized	overcame	perfected	pioneered
produced	recognized	reduced	served	simplified
solved	streamlined	taught	unified	wrote

Their tense and voice may be changed.

A list of objectives can be added, but be careful not to be too vague or limiting.

Never lie or exaggerate; this could lead to trouble. For example, overstating a proficiency in a foreign language can lead to embarrassment during the interview if the interviewer asks a question in that language. Avoid humor and flamboyant wording; use a simple, easy-to-read font. There is no need for visual material.

Resumé Hints

1. Assume the reader is intelligent.
2. A resumé is not a curriculum vitae or autobiography; keep it short.

3. The grammar of resumés is simplified: Subjects of sentences and personal pronouns are usually omitted, for example, *As part of a three-man team, decreased rejection rate 5 percent.*

4. Avoid the negative. Instead of saying *Almost met the targeted reduction rejection rate of 4 percent,* say *Achieved a 5 percent rejection decrease.*

5. Showing accomplishments is not boasting. Saving "something" for the interview is a mistake. Significant accomplishments should be included on the resumé to provide the greatest opportunity to be called for an interview.

6. List characteristics that are important to employers, including: leadership, organizational ability, good communication skills, problem solving, hard-working, and reliability.

7. Resumé templates seldom fit exact needs for specific jobs and should not be used.

8. Listing references is not required. At the bottom, state *References available on request.*

9. Proofread the resumé. It is a good idea to have someone else read the resumé before sending it.

Appearance

A resumé should be attractive and easy to read. It should fit on one page. Those with doctorates are exceptions; their resumés may be longer. Don't mix fonts; make sure the font and the size of the type are easily readable. Times New Roman is a good choice. Use only two sizes of type: one for the name and major headings and a smaller one for the rest. The margins should be one inch. Cut the number of words rather than shrinking the margins to squeeze in more words. Never staple a resumé; stapling makes it difficult to photocopy.

Example 1

Lloyd Bridges

Home address	2300 River Street,
	Hudson, NY 01234
	(231) 456-8910
School address	146 Oak Street,
	Rennselaer, NY 15678
	(423) 123-5678
email	bridge@RPI.edu

Experience:

Civil Construction Co., Syracuse,

NY (May–August 2007)
Leveling assistant in charge of ensuring level runways. Developed a simplified system of leveling.

Civil Engineering Dept., RPI, Rennsalaer,
NY (September 2006–May 2007)
Teaching assistant in CE 201, Strength of
Materials

Education:

Rennselaer Polytechnic Institute
B.S. Civil Engineering, expected May 2008
GPA 3.73
Relevant courses: Advanced Structural Analysis,
Concrete, Highway Construction
Project: Simulation of earthquake damage to
high-rise buildings

Skills:

CAD/CAM and other computer programs,
Weibull analyses, speaking knowledge
of Spanish

Honors:

3rd in class of 35. Dean's List (6 of 8 semesters)
Vice President, student ASCE chapter
(2007–2008)
Scholarship (2005–2008)

Activities:

Varsity ski team
Intramural football
Choir St Luke's Church in Albany
Served as Big Brother to young teen, 2007

Example 2

Cam Steel
202 Ford Street
Dearborn, MI 47172
(313) 145-6789
email: Steel@wsu.edu

Education
Wayne State University, B.S. Mechanical
  Engineering, expected 2009
GPA 3.36
Relevant courses: Strength of Materials, Physics,
  Computer Programming
Holy Cross High School, Dearborn, MI, 2005

Objective
Summer internship in automobile industry

Experience
Wayne State University Cafeteria (2006–2007),
  waiter
Joe's Auto Repair (2005–2007, after school),
  mechanic
High School debating team

Strengths
Hard-working, intelligent, organized, team player.
Willing to undertake any job.

Skills

Familiarity with UNIX, C, C++ computer
programs

Chapter Summary

Resumés should be short and direct. They should
emphasize the strengths of the candidate.

This concludes the formal presentation of
Reporting Results. The appendices that follow contain useful information on common errors in writing,
punctuation, and word choices to help develop writing skills. The appendices on the international system
of prefixes and units, on the Greek alphabet and its
typical uses, and on straight-line plotting of mathematical functions are provided as useful references in
technical writing.

Appendix I

Common Errors in Writing

This appendix is aimed at avoiding errors that the authors have seen in reading student papers and reviewing manuscripts for publication. Some of the examples are repeated from Chapter 1 for the convenience of the reader.

Pomposity

Avoid using large words where shorter ones would work just as well. For example, use *freezing* instead of *solidification*, *test* instead of *experimental investigation*, and *needs* instead of *requirements*.

Excessive Verbiage

A redundant word is an unnecessary word. Considering the high price of newsprint and book stock, we ought to watch for redundancies and pluck them from our writing as if we were picking ticks from a dog's back. Redundancies, like ticks, suck blood from our prose. Kilpatrick (1984)

Examples of excess verbiage are:

Example	*Instead use*
in order to	to
data points	data
at this point of time	now

Most uses of *respectively* can be eliminated without causing confusion. *Process* should be omitted in *casting process, machining process, rolling process*, etc.

Then can almost always be eliminated and should never be used more than once in the same paragraph. The following is a typical bad example: *The specimen was cut and then mounted in Bakelite. Then it was ground and polished.* Note that *The specimen was cut, mounted in Bakelite, ground, and polished* conveys the same meaning, but without the word *then*.

Avoid unnecessary redundancy. Instead of *try out, finish off, finally complete, absolutely necessary, triangular in shape*, and *very unique*, use *try, finish, complete, necessary, triangular,* and *unique.*

Apostrophes are also used in contractions to substitute for missing letters. For example, *The machine won't work* or *It's better to give than to receive.* Be careful to avoid the common mistake of confusing the possessive *its* with the contraction *it's*, meaning *it is*. Other common contractions are *can't, there's, haven't, we've,* and *he'd.* Although such contractions are acceptable in common speech and fiction writing, they should not be used in technical writing.

Quotation Marks

In American English, paired quotation marks (" ") are used for direct quotes, but not for paraphrasing of quotes. They can also be used in references around the title of an article or chapter. In addition, they can be used in text around a word or phrase that has questionable validity, as in *The professor said that an apostrophe was like a "bomb."*

In American English. for quotes within quotes, the inner quote is enclosed with single quote marks, as in *Mary said, "The professor quoted the book as saying 'Never assume that all equations are correct.'"* Note that in British English, the use of single and paired quote marks is reversed.

In American English, commas and periods always go inside quotation marks; the British put them outside unless they are part of the quote. Colons and semicolons always go outside quotation marks.

Colons and Semicolons

The rules for colons and semicolons are complicated. A colon is used as a mark of introduction when the clause, phrase, word, or series that follows the colon is linked to the preceding element. The most common use of a colon in technical writing is to introduce a list. For example, *An engineering decision must rest on a number of factors: short-term profitability, marketability, and safety.* Colons are also used in references to link subtitles to titles.

A colon may also introduce a phrase that explains, illustrates, amplifies, or restates the preceding phrase. For example, *Toughness is paramount in material choice for a pressure vessel: pressure vessels require high toughness.*

Semicolons are most often used to join two or more clauses when the second clause begins with a conjunctive adverb such as *accordingly, also, consequently, however, therefore,* or *thus,* as in the sentence *Stainless steel does not rust; therefore, it is used in the food industry, and it does not affect flavor.* However, the semicolon should not be used if the second clause is not closely related to the first. For example, the use of a semicolon in *Stainless steel finds application in the food industry; it contains at least 12 percent chromium* does not help explain why stainless steel is used in the food industry.

Semicolons can also be used to separate items in lists that contain internal commas. For example, *Metal Forming: Mechanics and Metallurgy*, third edition; *Materials Science: An Intermediate Text*; and *Materials for Engineers*, an undergraduate textbook.

Periods

Of course, a period indicates the end of a sentence. Periods are also used in abbreviations, such as *St. Venant's principle* and *et al.* Note that there is no period after *et* because it is a full Latin word meaning *and* but that *al.* is an abbreviation for *alia*, meaning *others.*

Italic Type

Italics are used in references to indicate a book title or a journal. Italics are used in mathematical expressions for variables. Note that in sin ($x/2$) the variable x is italicized, but the abbreviation of sine function (sin) and the number 2 are not.

Italics are also often used to introduce and explain a new word or phrase that may be unfamiliar to the reader. They may also be used in examples, as is done throughout this book. Italics may also be used for emphasis, but they should not be overused for this purpose.

Brackets

There are four types of brackets, namely: parentheses ( ), square brackets [ ], braces { }, and angle brackets < >.

Parentheses are used for explanatory words or comments, as in *Hill's first anisotropic yield criterion (1948) was of a quadratic form...*, *stretcher strains (also known as Lüders bands) are...*, and *work-hardening (strain-hardening)....*

In mathematical usage, angle brackets come outside of braces, which are outside of square brackets, which are outside of parentheses, that is, $\{ < [( )] > \}$ or $z = <\text{erf} \{\sin [1/(1-x)\}>^2$. The writer can sometimes simplify complicated expressions by breaking them into multiple equations. For example, the preceding equation written as $z = [\text{erf} (y)]^2$, where $y = \sin [1/(1-x)]$, is easier to read.

Ellipsis Points

An ellipsis, three consecutive periods,..., is used for trailing off, as in mathematical series like $x + x^2/2! + x^3/3! +...$, or to indicate missing words from a quotation.

Other Punctuation

Exclamation points should be avoided in technical writing. Question marks need no explanation, but they are rarely used in technical writing. Asterisks are sometimes used to designate footnotes. Bullets may be useful in oral presentations, but in the opinion of the authors have no place in technical writing.

Mathematical symbols, like $+$, $-$, $\neq$, $>$, $\geq$, $\leq$, and $/$, are used only in equations. The symbols @ and & should be avoided entirely, except in email addresses and in company names.

Capitalization

The first letter of the first word of a sentence is capitalized. A common mistake is to overcapitalize; within sentences only proper names should be capitalized. For example, *Poisson's ratio* is correct, not *Poisson's Ratio*. Names of elements are not capitalized, although the first letter of the symbol for a chemical element is. For example, *Cu* and *copper* are both correct.

Do not capitalize words following proper nouns. *Instron testing machine* is correct, but *Instron Testing Machine* is not.

For titles (figure titles, report titles, headings, etc.), capitalize either only the first word or all of the words except prepositions, articles, and conjunctions. For example, both *Schematic showing the drying concept* and *Schematic Showing the Drying Concept* are acceptable, but *Schematic showing the Drying Concept* is not.

Appendix III

Common Word Errors

There are many words that are easily confused with each other or that are commonly misspelled. Here is a collection that the authors find useful.

affect	v.t., produce an effect or influence
effect	v.t., to cause or accomplish; n., result or outcome
contaminate	v., to make impure
contaminant	n., a substance that contaminates
corroborate	v.t., to strengthen or confirm
collaborate	v.i., to cooperate or work with
discrete	adj., separate, disconnected
discreet	adj., tactful
ensure	v.t., to make sure something will happen
insure	v., to get protection

gage	n., test location of a tensile bar
gauge	n., instrument for measuring
grey / gray	grey is the British spelling of gray
plane	n., flat geometric surface; adj., flat, as in plane strain
plain	adj., simple or ordinary, as in plain carbon steel
principal	adj., first in rank, as in principal investigator or principal stress
principle	n., general truth or law
sample / specimen	a sample is a statistical group of specimens
silicon	n., element 28
silicone	n., a polymer with an S–O backbone
silica	n., SiO_2
stress	n., force per area
strength	n., critical value of stress, such as yield strength (not yield stress)

Spelling

It is *i* before *e* except after *c* and when sounding like *ay* as in *neighbor* or *weigh*. Many exceptions are contained in the sentence *The weird foreigner seizes neither leisure nor sport at its height.* Other exceptions include *either, being, obeisance, sheik, stein, counterfeit*, and *seismic*.

According to Henry Minott of the United Press, the fifteen most commonly misspelled words are:

changeable	dietitian	discernible	diphtheria
embarrass	gauge	harass	indispensable
judgment	likable	naphtha	occurred
paraphernalia	permissible	uncontrollable	

Other commonly misspelled words in technical writing are:

austenitizing	boundary	existence	foundry
height	inoculation	logarithm	martensite
regardless (not irregardless)	specimen	spheroidite	

Plural of Words of Greek or Latin Origin

Singular	Plural	Singular	Plural
analysis	analyses	appendix	appendices
colloquium	colloquia	criterion	criteria
datum	data	equilibrium	equilibria
focus	foci	index	indices
locus	loci	maximum	maxima
medium	media	minimum	minima
octahedron	octahedra	phenomenon	phenomena
tetrahedron	tetrahedra	thesis	theses
vacuum	vacua	vita	vitae

Use of Articles *a* and *an*

Whether *a* or *an* is used depends on the beginning sound of the following word or abbreviation. The article *a* is used before a consonant sound even if the word or abbreviation starts with a vowel. Examples are *a eutectic, a union, a U.S. senator, a one-time expense*, and *a UM professor*.

The article *an* is used before a noun or abbreviation that begins with a vowel sound even if the following word or abbreviation begins with a consonant. Examples are *an fcc lattice, an hour, an Rh factor, an n-p junction*, an MIT degree, *an unknown, an nth factorial*, and *an honor*.

Either *an* or *a* may be used before words that begin with a lightly stressed *h*. For example, *a history* and *an history* as well as *a heroic* and *an heroic* are acceptable.

Appendix IV

International System of Prefixes and Units

Table A.1. *Standard prefixes*

$10^3 n$	Name	Symbol
10^{-18}	atto	a
10^{-15}	femto	f
10^{-12}	pico	p
10^{-9}	nano	n
10^{-6}	micro	μ
10^{-3}	milli	m
10^3	kilo	k
10^6	mega	M
10^9	giga	G
10^{12}	tera	T

Table A.2. *Standard international system of units (SI)*

Symbol	Name	Quantity	Formula
A	ampere	electric current	base unit
Bq	becquerel	activity of a radio nuclide	1/s
C	coulomb	electric charge	A•s
°C	degree Celsius	temperature interval	base unit
cd	candela	luminous intensity	base unit
F	farad	electric capacitance	C/V
Gy	gray	absorbed dose	J/kg
g	gram	mass	kg/1000
H	Henry	inductance	Wb/A
Hz	Hertz	frequency	1/s
ha	hectare	area	$10{,}000 \text{ m}^2$
J	joule	energy, work, heat	N•m
K	Kelvin	temperature	base unit
kg	kilogram	mass	base unit
L	liter	volume	$\text{m}^3/1000$
lm	lumen	luminous flux	cd•sr
lx	lux	illuminance	lm/m^2
m	meter	length	base unit
mol	mole	amount of substance	base unit
N	Newton	force	kg•m/s^2
Pa	Pascal	pressure, stress	N/m^2
rad	radian	plane angle	(dimensionless)
S	Siemens	electric conductance	A/V
s	second	time	base unit
sr	steradian	solid angle	(dimensionless)
Sv	sievert	dose equivalent	J/kg
T	Tesla	magnetic flux density	Wb/m^2
t	tonne, metric ton	mass	1000 kg; Mg
V	volt	electric potential	W/A
W	ohm	electric resistance	V/A
W	watt	power, radiant flux	J/s
Wb	Weber	magnetic flux	V•s

Appendix V

The Greek Alphabet
and Typical Uses

Greek letters are frequently used for technical variables. The following table shows the most common usages for the Greek letters.

Letter	Symbol	Typical Use
alpha	A	
	α	angle, coefficient of thermal expansion
beta	B	
	β	angle
gamma	Γ	mathematical function
	γ	angle, shear strain, surface energy
delta	Δ	difference
	δ, ∂	difference between differential quantities
epsilon	E	
	ε	strain
zeta	Z	
	ζ	
eta	H	
	η	viscosity, efficiency

(continued)

Letter	Symbol	Typical Use
theta	Θ	temperature
	θ	angle
iota	I	
	ι	
kappa	K	
	κ	
lambda	Λ	
	λ	wavelength
mu	M	
	μ	coefficient of friction, 10^{-6}
nu	N	
	ν	frequency, Poisson's ratio
xi	Ξ	
	ξ	
omicron	O	
	o	
pi	Π	multiplicative series
	π	3.1415926…
rho	P	
	ρ	density, radius of curvature
sigma	Σ	summation
	σ	stress, conductivity, standard deviation
tau	T	
	τ	shear stress
upsilon	Υ	
	υ	
phi	Φ	
	ϕ	angle
chi	X	
	χ	
psi	Ψ	
	ψ	angle
omega	Ω	ohm, the end
	ω	angular frequency

Function	Plot	Graph
$y = a\exp(bx + cx^2)$	$\dfrac{\ln\,(y/y_1)}{x - x_1}$ vs. x	(graph: $\dfrac{\ln(y/y_1)}{x - x_1}$ vs x; Slope $= c$; intercept segment $b + cx_1$)
$y = a + bx + cx^2$	$\dfrac{y - y_1}{x - x_1}$ vs. x	(graph: $\dfrac{y - y_1}{x - x_1}$ vs x; Slope $= c$; intercept segment $b + cx_1$)
$y = \dfrac{a}{x} + b$	y vs. $\dfrac{1}{x}$	(graph: y vs $\frac{1}{x}$; Slope $= a$; intercept b)
$y = \dfrac{x}{a + bx}$	$\dfrac{1}{y}$ vs. $\dfrac{1}{x}$	(graph: $\frac{1}{y}$ vs $\frac{1}{x}$; Slope $= a$; intercept b)

Function	Plot	Graph
$y = \dfrac{x}{a + bx} + c$	$\dfrac{y - y_1}{x - x_1}$ vs. x	
$y = b + a\sqrt{x}$	y vs. $\sqrt{x}$ or y^2 vs. x	

For the first graph: vertical axis $\dfrac{y - y_1}{x - x_1}$, horizontal axis x, with Slope $= b + \dfrac{b^2}{a}x_1$ and intercept $a + bx_1$.

For the second graph: vertical axis y, horizontal axis $\sqrt{x}$, with Slope $= a$ and intercept b.

References

M. F. Ashby and D. R. H. Jones (1981), *Engineering Materials 1: An Introduction to Their Properties and Applications*, Butterworth/Heinemann, Oxford.

W. Chauvenet (1863/1961), "A Manual of Spherical and Practical Astronomy," Vol. II, *Theory of Astronomical Instruments: Method of Least Squares*, J. B. Lippincott, Philadelphia/Peter Smith Publisher, New York, pp. 564–66.

G. E. Dieter (1991), *Engineering Design: A Materials and Processing Approach*, second edition, McGraw-Hill Publishing Co., New York.

J. S. Dodd, ed. (1986), *The ACS Style Guide: A Manual for Authors and Editors*, American Chemical Society, Washington, D.C.

J. F. Hatch (1984), *Aluminum: Properties and Physical Metallurgy*, ASM, Metals Park, Ohio.

J. P. Holman (1989), *Experimental Methods for Engineers*, fifth edition, McGraw-Hill Book Co., New York, p. 63.

L. G. Johnson (1951), "The Median Ranks of Sample Values in Their Population with an Application to

Certain Fatigue Studies," *Industrial Mathematics*, vol. 2., pp. 1–9.

J. J. Kilpatrick (1984), *The Writer's Art*, Andrews, McMeel and Parker, Kansas City, Mo.

S. J. Kline and F. A. McClintock (1953), "Describing Uncertainties in Single-Sample Experiments," *Mechanical Engineering*, January, p. 3.

P. B. Schubert (1953), *Pipe and Tube Bending*, Industrial Press, New York.

L. Truss (2003), *Eats, Shoots and Leaves*, Profile Books, London.

Webster's Standard American Style Manual (1985), Merriam-Webster Inc., Springfield, Mass.

speaking, 43, 45
spelling, 134
standard deviation, 87
standard error of the mean, 91
straight-line plots, 141
stress vs. strength, 134
summary, 20, 28, 29, 43
systematic errors, 83

table of contents, 25
tables, 54
  format, 54
  numbering, 54
  titles, 54
technical data presentation, 53
technical letters
  action summary, 28
  examples, 29, 33
  format, 27
  organization, 26
  summary, 29
  text, 28
  three levels of presentation, 27
technical papers
  abstract, 17
  acknowledgments, 20
  discussion, 19
  experimental procedures, 19
  format, 13
  introduction, 18

references, 20
results, 19
summary or conclusions, 20
title, 17
tense
  past and perfect, 3
text of technical letters, 28
titles for tables, 54
titles of papers, 17
tone, 2

unambiguity, 7
uncertainty, 83, 86, 87
uncertainty analysis, 106

verbiage, 121
visual aids
  background and font color, 49, 51
  readability, 48
  text alignment, 50
voice, 3

walking the triangle, 44
Weibull statistics, 98
  characteristic value, 99
  lower bound, 100
  modulus, 99
  probability of occurrence, 100

Weibull statistics *(cont.)*
   scaling parameter, 99
   straight-line plots, 101
   three-parameter equation,
     101
   two-parameter equation, 101

word confusion, 133
wordiness, 121
   excess verbiage, 122
   process, 5, 122
   respectively, 122
   then, 11, 122